50 Ways to Kill a Slug

An Hachette UK Company
www.hachette.co.uk

First published in 2003 by
Hamlyn, a division of Octopus Publishing Group Ltd.
Endeavour House, 189 Shaftesbury Avenue,
London WC2H 8JY
www.octopusbooks.co.uk
www.octopusbooksusa.com

Distributed in the U.S. and Canada by Octopus
Books USA:
c/o Hachette Book Group USA
237 Park Avenue
New York NY 10017

ISBN: 978-0-600-60858-5

A CIP catalogue record for this book
is available from the British Library

Printed and bound in China

20 19 18 17 16

50 Ways to Kill a Slug

Sarah Ford

hamlyn

For Mum, Dad and Matt

NOTES

All recipes serve four unless otherwise stated.

Slugs should be cooked thoroughly. To test if a slug is cooked, pierce the flesh through the thickest part with a skewer or fork – the juices should run clear, never pink or red.

A few recipes include slugs. Anyone with a known slug allergy should avoid these.

While all reasonable care has been taken during the preparation of this book, neither the publisher, editors or author can accept responsibility for any consequences (such as a revenge attack by the slugs) arising from the use of this information.

Always read the manufacturer's instructions carefully before using any chemical or organic pesticide. Keep them away from animals and children, but wield gleefully in front of slugs.

Contents

INTRODUCTION

KNOW YOUR ENEMY

After a long winter, there's nothing like seeing your garden spring back to life. However, newly formed shoots, blossoming flowers and moist soil are heaven for the slug. Not only can slugs munch through all your seedlings in minutes, but they can also destroy your favourite hosta in a matter of days.

HISTORICALLY, BATTLES HAVE BEEN LOST BECAUSE...

The slug is a hermaphrodite, which means it has luck on its side when it comes to mustering troops for combat. Hermaphrodites have both male and female reproductive systems, so slugs can mate with themselves when and where they wish – although it has been known for a slug to choose to make love with another slug for up to 90 minutes! As a consequence of these activities, each slug can produce up to 36 eggs, several times a year, which is a huge number of new slugs. The white-jelly eggs are laid underground and hatch within 10–21 days, and the baby slugs reach adulthood in approximately six weeks.

A slug has a life span of several years, and every year it will grow bigger and destroy more of your garden. An average garden contains approximately 200 slugs, and in a season each slug can eat up to 0.8 kg (1¾ lb) of plants. Consequently, you could be forgiven for thinking that you are fighting a losing battle!

THEIR STRENGTHS AND WEAKNESSES

Slugs shelter by day in dark, moist areas and come to life by night, making them difficult for humans to find, let alone fight. However, all is not lost as there are certain things that a slug just cannot do without. Originally the slug was a sea creature: its gills became lungs so that it could survive on land, but it still needs moisture as this is constantly lost through the slime it produces when it moves. To produce slime, slugs have to eat and drink constantly. This knowledge is your first weapon in beating your enemy.

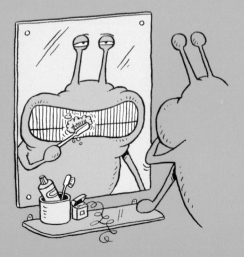

ENEMY PROFILE

NAME:	Slug (*Arion lusitanicus*)
SIZE:	0–15 cm (0–6 in)
SPEED:	0.0113 kmph (0.007 mph)
REPRODUCTION:	Hermaphrodite – produces over 100 eggs a year
LIFESPAN:	2–6 years
HABITAT:	Dark, moist shelters with a constant supply of tasty plants
HABITS:	Destroying plants, leaving slime trails and eating worms and, worse, dead slugs
WEAPONS:	An ability to camouflage and attack under cover of darkness, an insatiable appetite (can eat double its own body weight in a day) and 27,000 teeth
LIKES:	Beer, grapefruit rind and cabbage
DISLIKES:	Salt
ENEMIES:	Birds, frogs and badgers
PREY:	Succulent green leaves (particularly delphiniums and hostas), and they have a penchant for the skins of citrus fruit (*see page* 46)

PREPARE THE BATTLEGROUND

There are many approaches you can take to minimize the chances of slugs thriving in your garden. Start by eliminating their favourite habitats.

1. Prune shrub branches that touch the ground.

2. Trim grass edges, as slugs hide in the overhang.

3. Remove excess mulch.

4. Check out their favourite hideaways – under decking, around compost heaps, anywhere dark and moist.

5. Spring-clean your garden regularly, and hoe your weeds as slugs and eggs hide under them, as well as under earth clods.

6. Encourage slug predators – many creatures are known to feast on the odd slug or two, including moles, hedgehogs, frogs, salamanders, carnivorous beetles and millipedes.

7. Plant tender seedlings somewhere safe, away from the prying eye of the slug, or protect them by surrounding them with the cardboard centres of toilet rolls.

GATHER YOUR WEAPONS

With the following implements at your disposal
you will be ready to take up arms and commence
battle at a moment's notice!

SQUIRTY SPRAY GUN Fill this with all manner of
liquids that are distasteful to slugs (*see page* 37)
and take aim.

CHOPSTICKS If handpicking is not your thing and
you prefer the oriental approach, then flick them
off with these handy implements. (Don't forget just
exactly what you are doing and make the mistake
of thinking you are eating sushi.)

FLASHLIGHT OR TORCH Essential for exposing the
enemy at night.

CATAPULT See how far your captives will go
when catapulted off the edge of a cliff. Mean-
spirited individuals may want to aim them at
next-door's hosta.

RUBBER GLOVES For those of you who have an
aversion to slime, these are very useful for
handling slugs.

BUCKET A suitable container for collecting,
drowning and cleaning up.

POKER OR SHARP STICK For wielding, scaring and stabbing – at arm's length, of course.

BEER For getting slugs drunk and drowning them, and for a celebratory drink after you've defeated the slug. (Don't drink the same beer in which you drowned the slug!)

SALT Great for fish and chips, not so good for slugs.

HOE Sort out those weedy slug hideouts and tidy your garden at the same time.

PREDATORS A few of these are very handy (*see page 55*), although not always very practical in a small city garden – they can play havoc with the neighbours.

BARRIERS AND BAITS A full list of these is given on pages 24 and 50, so arm yourself and be victorious.

VINEGAR A good ingredient for slug sprays and very useful for removing slug slime.

WARM SOAPY WATER Great for drowning slugs and also washing your hands.

SPADE Guaranteed to chop even the toughest slug in half.

SWEETS AND MONEY Necessary to bribe your children and their friends.

WHAT TO DO WHEN YOU'VE BEEN SLIMED

Slug slime is grim. However, in your wars against slugs, it is inevitable that you will get slimed on more than one occasion. So, to de-slime, mix a little vinegar with warm water and wash your hands. Please be careful to avoid any cuts on your hands.

WHAT TO DO WITH A SLUG (DEAD OR ALIVE)

You've killed and collected them, or just collected them. What now?

Slugs are much more useful than you might imagine. They are full of protein and can make a tasty snack for cats, ducks or fish. They are also quite a good addition to a compost bin: as the average compost pile is full of their favourite food, the living ones will love you forever and will stay put; they will also help with the breakdown of cardboard and paper and improve the quality of your compost. With good-quality compost, your plants will become stronger giving them more of a fighting chance against the slug! The dead slugs will be eaten by the living slugs, who in turn will become strong, making better compost – and so the cycle continues.

ENTER INTO COMBAT...

50 WAYS TO BEAT YOUR ENEMY

PREVENTION

KEEP YOUR GARDEN TIDY

One of the main reasons that gardens become slug-infested is because of all the decaying matter that builds up over time. It's time for a bit of spring-cleaning!

1. Dedicate an afternoon to clearing out all potential slug hideouts, under decking, around compost heaps and in any other secluded, dark, moist areas.

2. Clear your garden, systematically removing all decaying matter from beds and borders. Do this on a regular basis and add anything suitable to your compost pile, along with any slugs you find (*see page* 17). Make sure your compost pile is located away from growing areas.

3. Remember, a tidy garden is a slug-free garden.

CONFUSE-A-SLUG

It isn't always easy to do this, but it's worth the effort. Slugs, like good scouts, will follow their friends' tracks, knowing that these head towards food.

1. Check your garden for the glint of slime.

2. Once you have found a slime trail, destroy this fast-track to the bounty: take up your trowel and dig it over.

3. Survey your handiwork.

4. Wash your trowel and your hands with warm water and vinegar to get rid of excess slime.

DON'T LET SLEEPING SLUGS LIE

If you catch them when they're snoozing, they're sitting ducks!

1. Aim to cultivate your soil in the spring when slugs are in hibernation. This way you'll kill them during their long winter nap and they won't know a thing about it.

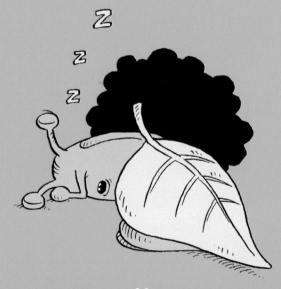

MAKE SLUG CAVIAR

A good organic way to beat the blighters! The eggs
are small and white and are laid all year round.

1. Gently hoe your garden to expose slug eggs.

2. Stand back and allow predators, such as
hungry birds, to eat the rich pickings –
they'll have a real treat. Your garden will
look great, too.

BARRIERS

LET PLANTS FIGHT BACK

There are certain plants that slugs hate.

1. Choose the following from your local garden centre: mint, chives, garlic, geraniums, foxgloves and fennel.

2. Plant a selection around the edge of your garden. Slugs really dislike the strong smell of these plants, which means they'll act as a barrier and stop gastropod infiltration.

WREAK PREHISTORIC REVENGE

It's ironic that all these years later some plants can get their own back. (Diatomaceous earth is non-toxic but try to avoid inhaling the dust. It's not available everywhere, but is worth seeking out.)

1. Obtain some diatomaceous earth. This is a powder made from the ground skeletons of small, fossilized plants. The small razor-sharp edges slice up the slugs when they try to slime across it, causing them to dehydrate and die.

2. Dust your plant and the surrounding soil with the diatomaceous earth.

EMPLOY SOOTY DE-SLUG

Soot is a great barrier – slugs detest it! It acts by drying up the slug's mucous glands and, therefore, restricting its movement. If it's mixed with ashes and lime, it will be even more effective as these contain other chemicals that slugs dislike. Mixed together, the ingredients make a winning team.

1. Scrape some soot out of your chimney and sprinkle it around the base of your plants.

2. Reapply after rain!

3. Apply some extra soot to your face and do a war dance.

GIVE THEM (MORE THAN) BLISTERS

Eggshells, grit and sand are great slug barriers!
Other successful barriers are crushed nuts,
seashells, gravel and pine needles. So, if at first you
don't succeed, try something else. A bit like you
getting a stone in your shoe, the slug will
experience great discomfort if it tries to slime
over this little lot.

1. Sprinkle a mixture of the barrier ingredients around the base of your plant and watch the slugs as they look longingly up at a plant they'll never get to nibble.

2. Hang signs around your plants saying, 'Come and get me, if you think you're hard enough!' Guaranteed to increase the slugs' frustration.

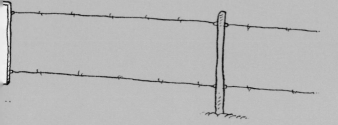

MAKE TIME FOR A HAIRCUT

If you suffer from a lack of hair on your head, then perhaps a neighbour could oblige, or a pet dog or cat. Hair makes a great barrier. No matter how silky your locks, a slug will be reluctant to cross them.

1. Instead of throwing away cut hair, collect it and sprinkle it around the base of your plants.

2. The coarse texture of the hair will stop slugs getting through.

3. Save some to reapply regularly.

31

USE DIY SLUG DEFENCE

Never fear. All the sawdust produced during a day
of DIY need not be wasted!

1. Gather sawdust from under your workbench.

2. Sprinkle it around the base of your plants.
This is a proven barrier, too rough
for even the bravest of slugs to attempt
to cross.

ARRANGE A LANDSLIDE

Vermiculite is normally used to aerate and loosen
soil and potting compost; it also works as an
effective slug barrier.

1. Add water to the vermiculite so it swells up.

2. Place this mixture around your plants.

3. Much as the slugs are attracted to the moisture,
they will not like the movement as for them
it will feel like a landslide is about to take place.

SEND THEM OFF-PISTE

This is a great way to watch a slug on
the slippery slope.

1. Put petroleum jelly around the base and tops
of your plant containers.

2. Watch Mr Slug having to muster all his
energies to get past this gummy
barrier...it's stick or slip for the slug!

GET THEM TAPED

Tapes containing chemicals are readily available in garden centres. They act as an effective barrier around a plant and are a permanent end to a slug's chomping days.

1. Apply anti-slug tapes on the soil surface or in the ground around each plant.

2. Hold the tape in place with stones or earth. When slugs and snails chew it, they suffer fatal indigestion.

3. From the holes made in the tape, it is possible to pinpoint the areas that are most infested and to use additional tape to eliminate the pests.

APPLY EAU DE YUCCA

Usually found in an aerosol spray, extract of yucca is repugnant to slugs. Alternatively, make up a spray yourself to protect your plants...all those suggested are a turn-off for slugs!

1. Stick a yucca leaf in a food processor, process, then mix with water, fill up your spray gun and blast your plants.

2. Other mixtures that will work are garlic in liquid paraffin; vinegar with water; a mixture of boiling water, chillies, an inch of chopped horseradish root and a couple of handfuls of geranium leaves. When cool, strain into your spray gun (use only one mixture at a time) and spray the garden.

RESORT TO SHOCK TACTICS

A short sharp shock is sometimes all it takes to keep gastropods at bay.

1. Most garden centres stock self-adhesive copper tape that can be wrapped around the base of a plant pot. This creates a physical barrier that slugs won't cross.

2. The copper also naturally carries a tiny positive electric charge, and when the slugs crawl up the container, they are given a small shock, which soon sends them scooting back down.

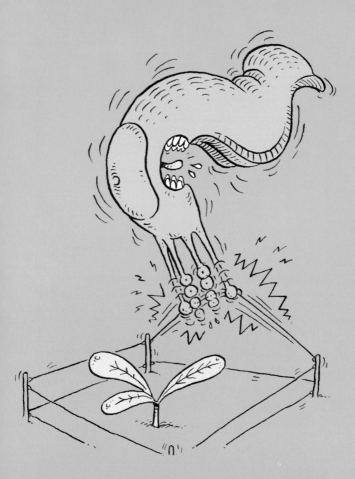

TRAPS

GET THEM DEAD DRUNK

For those of you whose karma may be rocked by slug-slaying, this is the method for you...what nicer way for a slug to go?

1. Fill a shallow bowl with beer (Michelob and Budweiser are thought to have a high success rate) and place it strategically in an area of high slug activity.

2. Wait overnight.

3. Slugs love a slurp of stale beer. Once in it, they'll drown.

4. Disposing of the slugged brew can be a messy business. A simple solution is to add it to your compost heap (*see page* 17).

GIVE THEM A BUZZ

Recent research has proven that caffeine has the power to stop slugs in their tracks! However, it is not known whether it can also harm other more innocent garden dwellers and it is not yet fully licensed as a slug killer.

1. Fill a shallow container with some coffee or cola and place on the edge of your border, close to your prize plants.

2. Watch the slugs take that coffee-break, never to return!

43

PREPARE A GOURMET MEAL

For those of you who really don't want to kill slugs but do want to protect your plants – why not distract them with this, the nicest of slug traps?

1. Place either cabbage or lettuce leaves upside down among plants that have been munched.

2. Slugs will congregate under the leafy shells and eat away. If you don't want to kill them just relocate the slugs a safe distance from your plants. However, don't think it's all over: both slugs and snails have an inbuilt homing device – they will return!

CREATE A SLUG PLAYHOUSE

Buying a grapefruit at the supermarket is a double investment: not only can you enjoy it for your breakfast, but you can also use the skin as an organic trap or playhouse for slugs – they just love grapefruit!

1. Halve your grapefruit and eat it for breakfast.

2. Make a small hole for slugs to enter the grapefruit skin. Place the skins upside down around the garden and wait a couple of days.

3. Your plants will be left alone as the slugs huddle together under the skin and celebrate their delicious find.

4. Collect up the slugs and dispose of them.

WEAVE A MAGIC CARPET

The next time you are taking a worn-out carpet to
the tip, save a few small pieces to make a slug trap.

1. Dampen your carpet scraps with water and
spread them randomly around your garden
among your plants.

2. Within days slugs will be drawn to the carpet
like bees to a honey pot.

3. Scrape off the slugs and dispose of them
in your compost bin (*see page* 17). Reuse your
carpet pieces until you can't bear to touch
them anymore.

BAITS

LAY ON THEIR LAST BREAKFAST

Good for us, not so good for them!

1. Scatter oat bran around your plants and wait for the slugs to eat it for breakfast.

2. Oat bran is a favourite with slugs, but once eaten it expands inside them and their end is nigh – they have been known to literally explode.

3. If all the bran has been used up killing slugs, treat yourself to a blueberry muffin for breakfast, instead.

LET THEM EAT CAT FOOD

Slugs love cat and dog food, and a ready supply
will coax them away from your plants. The food
will also attract hedgehogs, which will eat slugs – a
double whammy! On the downside, it could have
the adverse effect of encouraging visits from all the
local cats and dogs – turning your garden into a
giant litter tray.

1. Bury some nearly empty cans of cat or dog food around the slug-infested areas of your garden. Temptation should be too great and, hopefully, the slugs will fall in trying to get to the feast.

2. Pick out and destroy or otherwise dispose of the victims.

ORGANIC IDEAS

CALL IN THE PREDATORS

Build a wildlife garden that is attractive to slug
eaters – thrushes, frogs, toads, beetles and
hedgehogs are all rather partial to a juicy
slug or two.

1. Dig a pond for frogs and toads.

2. Position a bird bath and put food on
a bird table.

3. Leave out hedgehog fodder.

4. Wait for the influx of the right kind of
wildlife, which will reduce your slug
population without guilt.

CON THEM WITH COMFREY

Sacrifice a few comfrey plants for the sake of all the other plant inhabitants in your garden...the greatest good for the greatest number!

1. Slugs love comfrey! Plant them in particular slug hotspots.

2. Within a couple of days these poor martyrs will be covered with slugs. Pick them off and wait for the next batch. Dispose of as you see fit.

EMPTY THEIR LARDER

If all else fails, grow a garden full of plants that
slugs dislike (*see also page* 25). Not as much
fun as some of the other solutions but
successful nonetheless.

1. Grow some of the following plants: artemesia,
bleeding heart, cornflower, forget-me-not,
fuchsia, hydrangea, lavender, nasturtium, peony,
pink, tulip and wallflower.

2. It is debatable whether there is such a thing
as a plant that can withstand the voracious
appetite of the slug, however, these will fare better
than hostas or delphiniums, which are its caviar.

ARRANGE A RELOCATION, RELOCATION, RELOCATION

One of the kinder ways to protect your garden from the slugs' insatiable appetite is to force them to move house!

1. Under cover of night, take up your torch or flashlight and your bucket and venture out into the garden.

2. Using the element of surprise, flash your torch around all the usual slug haunts (your hostas and the vegetable patch). Collect the slimy creatures in your bucket.

3. Drive to your nearest woodland and empty your bucket. Slugs like nothing more than damp woods, ferns, moss and lichens! You'll just have to hope that word doesn't get out among the slug fraternity, otherwise they'll all head for your garden hoping to be upgraded to their dream wood!

DARE THEM INTO THE CHICKEN RUN

Chickens are a great investment if your garden is big and rural: not only will they clear your garden of gastropods, but they'll also ensure that you have a regular supply of eggs!

1. Choose three hens and one cock (Rhode Island Reds are your best option, as they have an insatiable appetite for slugs), so your egg supply is guaranteed.

2. Set up a heavenly chicken coop at the bottom of the garden.

3. Let the chickens loose to do their thing and munch on all the slugs – they'll need little encouragement!

4. Enjoy scrambled eggs for breakfast in the knowledge that your garden is slug-free and your plants are beautiful.

CHEMICAL
CURE-ALLS

POP A PELLET

Metaldehyde pellets are guaranteed to kill slugs.
However, there is some controversy over their
safety. Always store them responsibly, away
from animals and children and follow the
manufacturer's instructions when using them.

1. Distribute pellets thinly and evenly over beds
and borders and wait for the slugs to chomp
away on their favourite blue sweets. The best
time to use them is early evening and on wet
or damp days.

2. If you are concerned about animals or children
getting hold of the pellets, cut a plastic bottle
in half. Bury it in the soil with the neck end
just peeking out. Feed pellets into the opening.
This will ensure that only the slugs can get
to them.

SPRAY SLUG-SLAY

Chemical sprays are not always popular but can be
highly effective.

1. Some metaldehyde- and methiocarb-based
products come as aerosols. Follow the
manufacturer's instructions, which usually
involve spraying the soil: slugs and snails are
destroyed on their nightly excursions when
they come into contact with the sprayed earth.

2. Pick up and dispose of the dead slugs so
that other, more precious wildlife can't
get poisoned by them.

MAKE WAY FOR A SPRAY

This is a safer spray or powder but not as effective.
It kills slugs by absorbing water from their slime-
producing organs, but can also damage some
plants, so be careful when spraying.

1. Mix 2 teaspoons of iron sulphate with
4 pints of water (or purchase a spray
from the garden centre).

2. Spray your slug-infested areas.

3. Reapply regularly and clear up the victims.

BE THE FASTEST GUN IN THE GARDEN

Use these tactics to search out and destroy!

1. Fill a squirty spray gun with 1 part ammonia to 3 parts water.

2. Head into the garden in the dead of night with a torch or flashlight and seek out your enemy.

3. Just one squirt and the slug will be history.

PASS THE SALT

Salt is a highly effective way to kill slugs, and one particularly favoured by young boys, but it can have some negative consequences for your garden, so use it sparingly.

1. Under cover of night, arm yourself with a torch or flashlight and a handy pack of table salt.

2. Sprinkle the offending mollusc with just a few grains of salt.

3. Watch joyfully as it curls itself into a little ball; it will be dead within seconds.

USE A PURGATIVE

This is a good, but perhaps somewhat impractical, way of killing the critters.

1. Just sprinkle magnesium salts over the soil around your plants.

2. Repeat applications after rain or watering. The salts will burn the slugs and discourage them from attacking the plants.

3. Alternatively, sprinkle the salts directly on to the slugs and watch them shrivel.

BIOLOGICAL
ALLIES

NOBBLE THEM WITH NEMATODES

By using nematodes you can control slugs safely, effectively and organically. The packs of powder contain millions of microscopic nematodes (parasites) that kill slugs both above and below ground.

1. Simply mix the powder with water and apply to the soil using a watering can.

2. If you follow the manufacturer's instructions the nematodes should remain active for six weeks, even during prolonged bouts of wet weather, when slugs are at their most rampant.

BEFRIEND A
SLUG-EATING PLANT

Dutch scientists have produced a rare slug-eating
carnivorous triffid. This predator plant could
be the next big thing in the battle between
man and slug.

1. Start saving your money as this pioneer
plant is unlikely to come cheap, although
research is continuing to see if it can be
mass-produced and survive in the wild.

SIMPLY SILLY

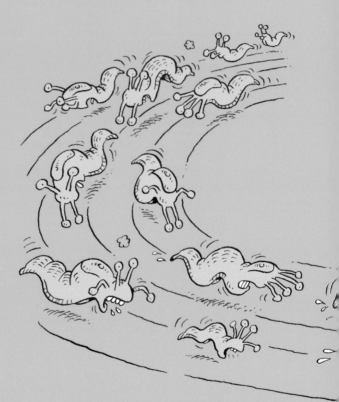

HOLD A SLUG RACE

A great way to dispose of several slugs at once –
at very little cost.

1. Locate a small child.

2. Talk him and several friends into collecting
slugs for you on the guise that you are
holding a slug race.

3. Give a bar of chocolate to the child whose
slug won.

4. Gather up the slugs and dispose of them.

THROW A SLUG-SLAYING PARTY

For the more bloodthirsty slug-slaughterer – this will be more successful and more fun by night.

1. Invite your friends to a slug-killing party in your garden. You can really enter into the spirit of things by dressing up.

2. Provide everyone with torches or flashlights and spades. Locate the slugs with the torches and chop 'em in half with your spade.

3. Offer a prize for the person who kills the most gastropods.

4. Celebrate the death of the slug and the new life of your plants!

HANG THEM OUT TO DRY

A slightly time-consuming but amusing way to
dry out slugs.

1. Collect a bucketful of slugs.

2. Wearing rubber gloves, peg each one to your
washing line.

3. Watch as they wriggle and sway gently in the
breeze and dry out and die in the sun.

FLAP THEM TO DEATH

Slugs love dampness. One way to make their environment increasingly unpleasant is to increase air movement and, therefore, reduce moisture! This is probably not one of the easier options as you'll have to flap your arms for some time before you have any effect. However, one consolation is that you'll become very fit in the process.

1. Dress in your most sporty outfit.

2. Run around in your garden and flap your arms about your head wildly.

3. Continue this activity for some time.

4. With some increased air movement there is an outside chance that the slug might choose to move on to a moister garden.

PLAY HIDE AND SEEK

Play this with your slug enemies on a weekly basis. Unfortunately, you will always be cast in the role of the seeker, which can become a bit tiresome, but it will pay off.

1. Slugs are crafty creatures and by day will hide in tree trunks, weedy patches, under stones and decking. Try to get to know all the places where they might be taking a snooze.

2. Physically remove them on a regular basis and dispose of as you see fit.

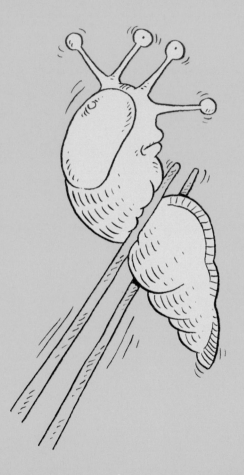

MAKE SLUG-SUEY

One of the best ways to get rid of slugs is to pick them off your plants by hand. While this is a boring and unappetizing task, it is highly effective. If, however, you cannot face the thought of coming into contact with the slug, try the Chinese method.

1. Master the fine art of using chopsticks.

2. Brandishing your chopsticks over your devoured plants, remove each slug with a quick flick of the wrist. This may make the whole process slightly less disgusting. However, your next-door neighbours won't thank you if the slug ends up in their garden, so practise your aim first.

IMPRESS THEM WITH YOUR TREADS

This one is only for those with strong stomachs –
it may not be the kindest way to send slugs off,
but it's very satisfying.

1. Tell your children that you want to give them
a biology lesson and get them out into the
garden in search of slugs.

2. When they have successfully collected enough
of the little beasts, give them a quick talk –
dipping into your wealth of knowledge about
slugs – then send them to bed.

3. Once they are safely tucked up, wreak
your revenge by spreading all those slugs on your
driveway and driving back and forth over them
in your car.

SEND THEM INTO ORBIT

Although only a temporary method of losing the slug, this is a good way to practise your throwing skills, and one that can provide hours of amusement if you invite a few friends around and have a competition to see whose slug goes the furthest.

1. Find your slug.

2. Take aim.

3. Throw.

4. Shout after it, 'Don't come back again!'

SPIN A SLUG SMOOTHIE

Slugs make nutritious snacks for certain animals (fish, cats, ducks) as they contain protein, vitamins, calcium and oils.

1. Handpick a selection of slugs.

2. When you have enough, stick them in a blender and do a quick whizz to make a lovely slimy brew. This delicious and highly nutritious smoothie will benefit your cat (and garden) no end.

3. Alternatively, kill the slugs first and feed them to your fish, which will merrily snack away on them.

4. Don't forget to give your blender a thorough clean!

85

DO THE SLUG SHUFFLE

Keep fit and reduce the chances of slug survival by doing a weather-altering dance. During periods of intense heat and sunshine slugs dry out and shrivel or remain buried underground. Similarly, during heavy rain they'll keep a low profile.

1. Do your best raindance.

2. If this doesn't work or you get fed up with rain, try a sundance instead. Either way you'll drive them underground!

TORCH THEM! SCORCH THEM!

The really mean among you will take great delight in scorching the little blighters with a weed-burner.

1. Gather some slugs together in a group.

2. Light your torch and advance on the enemy. The bloody attack will be over with one quick blast from the torch and you will be the winner.

RELY ON A RED HOT POKER

For the bloodthirsty – instant slug death!

1. Ensure that your battleground has been
prepared with pokers randomly spaced
around the garden.

2. Eye up your enemy, then face him –
man to slug.

3. Wield your poker and impale him. Once dead,
flick him triumphantly over the hedge
(checking first for passers-by) in full view of
all his slug friends.

4. Wash your poker.

GIVE THEM A BATH

A simple solution that will send the slug to its grave smelling sweet.

1. Fill a bucket with warm soapy water.

2. Venture out into the night disguised as a private slug sleuth. Collect as many as possible and drown them in the bucket. Show no mercy, after all they don't show any to your plants!

SEND THEM ON A LONG TRIP OFF A HIGH CLIFF!

Create a new sport with the help of your adversaries, the slugs. To be effective and fun, you'll need to collect quite a few slugs over a period of time.

1. With a full bucket of slugs, head for the nearest cliff, along with a few friends and a selection of catapults.

2. Take up a position on the edge of the cliff (not too close to the edge).

3. Load your catapult with slugs and send them skydiving, never to return.

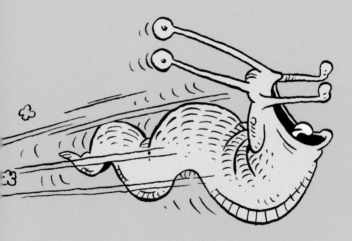

IF ALL ELSE FAILS...

Venture out into the garden armed with this book.
Locate a slug, remove it from among your plants,
place the book unopened on top of the slug and
squelch down with your foot. Then flick off the
dead remains. Finally, wipe down your book!

95

THE END

ACKNOWLEDGEMENTS

Executive Editor: Sarah Ford
Editor: Alice Tyler
Executive Art Editor: Geoff Fennell
Illustrator: KJA-artists.com
Production Controller: Sam Coleman